AF871600

# What a Qubit Actually Is

Quantum Computing, Interference, Entanglement, and What the Science Really Says About the Multiverse

Kristian Magnuson

ISBN 979-8-9976881-3-4

## Contents

# What a Qubit Actually Is

### Quantum Computing, Interference, Entanglement, and What the Science Really Says About the Multiverse

Kristian Magnuson
OpX Advisory Group

Second Edition - 22 August 2026
Paperback ISBN 979-8-9976881-3-4

Sources current to August 2026. Vendor claims are labelled as such throughout.
Verification caveats are listed before the references.

---

## Abstract

This paper examines the physical mechanism of quantum computation at the level of the underlying physics, and then applies that mechanism to a question frequently raised in popular accounts: whether the operation of quantum computers constitutes evidence for the existence of parallel universes.

The two halves of the paper are connected by a single technical fact. The popular claim that a quantum computer "tries all possible answers simultaneously" is not a simplification of the truth; it is false, and provably so. Quantum algorithms do not evaluate every candidate in parallel and select the winner. They arrange interference — the cancellation of amplitude on wrong answers and its reinforcement on right ones. This distinction is not pedantic. It is the difference between a machine that would solve NP-complete problems instantly (which no quantum computer can do, by theorem) and one that solves a narrow, structured class of problems (which is what we have built).

That correction turns out to be decisive for the multiverse question. The intuition that quantum computers demonstrate parallel universes rests almost entirely on the parallelism picture. Once the mechanism is stated correctly, the argument requires substantial repair — and the leading defender of the multiverse reading, David Deutsch, has himself abandoned the "collection of parallel classical universes" framing that makes the popular version compelling.

We conclude that quantum computation is interpretation-neutral: it works identically, and is explained adequately, under every mainstream interpretation of quantum mechanics. It is not evidence for the Many-Worlds Interpretation. It is also not evidence against it. What it is — and this is more interesting than the usual framing allows — is a genuine and tightening experimental constraint on a rival class of theories in which the wavefunction physically collapses.

---

## How to read this paper

This paper has four parts and they are not equally technical.

**Part I** builds the object. What a qubit is, why amplitudes rather than probabilities, what superposition does and does not mean, what measurement destroys, and why decoherence makes the engineering hard.

**Part II** is the argument. It establishes, with theorems rather than assertions, that quantum speedup comes from interference and not from parallelism.

**Part III** is the hardware. Where the field actually stands, separated carefully from where vendors say it stands.

**Part IV** applies Parts I and II to the multiverse question.

**If you read only two parts, read II and IV.** Part II contains the correction that the rest of the paper depends on; Part IV is where it pays off.

### The evidence standard

Part III in particular reports claims of very different quality. A peer-reviewed result independently benchmarked by a national laboratory and a number in a press release are not the same kind of fact, and this paper refuses to present them as though they were. Every load-bearing empirical claim about hardware carries a tag:

| Grade | Meaning |
|---|---|
| **[V] Verified** | Primary source: peer-reviewed publication, national-laboratory benchmark, or a mathematical theorem with a citable proof |
| **[R] Reported** | Credible trade press or secondary analysis, not independently confirmed |
| **[P] Projected** | Forward-looking vendor, analyst, or roadmap statement. Not an observed result |
| **[A] Assumption** | The author's own judgment, flagged so you can disagree with it separately from the facts underneath |

The habit worth taking away is the diagnostic question. When someone tells you a quantum computer will do something by a given year, ask which tag belongs on that sentence. In this field the answer is almost always **[P]**, and almost nobody says so.

Theorems are graded **[V]** throughout. A proof is the strongest evidence available in this subject, and most of the load in Parts I and II is carried by proofs rather than by experiments.

---

**List of figures**



---

# Part I — The Physics

## 1. The classical bit, and the shape of its state space

A classical bit is an element of a two-element set, $\{0, 1\}$. That sentence contains everything there is to say about it. A bit is in one of two states; if we are uncertain which, we can describe our ignorance with a probability $p$ of being 0 and $1 - p$ of being 1, but the bit itself is always definitely one or the other. Our uncertainty is a fact about us, not about the bit.

It helps to picture the state spaces geometrically, because the comparison to a qubit becomes immediate.

- A **deterministic classical bit** has a state space of two points.
- A **probabilistic classical bit** has a state space that is the line segment between them — the set of all $(p, 1-p)$. One continuous parameter.
- A **qubit**, as we will see, has a state space that is the *surface of a sphere*.

The extra dimension in that progression is not a metaphor. It is where every non-classical effect in this paper comes from.

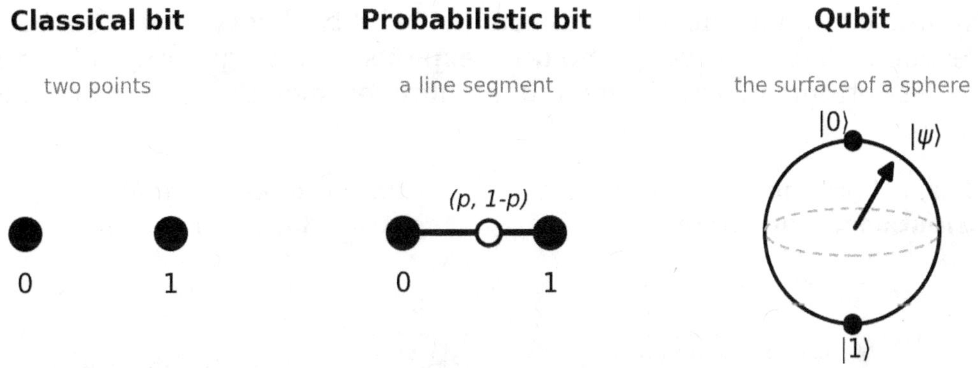


***Figure 1. Three state spaces, drawn to the same scale of ambition.*** *The progression is not rhetorical. A classical bit occupies one of two points. Allowing ignorance stretches that to a line. A qubit's state space is the entire surface of a sphere — and the extra dimension, the circle of relative phase, is where every non-classical effect in quantum computing originates.*

## 2. What a qubit is

### 2.1 The state vector

A qubit is a unit vector in a two-dimensional complex vector space (a Hilbert space, $\mathcal{H} \cong \mathbb{C}^2$), considered up to an overall phase factor:

$$|\psi\rangle = \alpha|0\rangle + \beta|1\rangle \qquad \alpha, \beta \in \mathbb{C} \qquad |\alpha|^2 + |\beta|^2 = 1$$

The numbers $\alpha$ and $\beta$ are called **probability amplitudes**. They are complex numbers. This is the single most important structural fact in quantum computing and it is worth dwelling on before going further.

### 2.2 Why amplitudes and not probabilities

Probabilities are non-negative real numbers, and they add. If there are two independent ways for an event to happen, each with probability $\frac{1}{4}$, the total probability is $\frac{1}{2}$. Probabilities accumulate; they never destroy one another.

Amplitudes can be negative. They can be complex. And therefore **they can cancel**.

Suppose two computational paths both lead to the same outcome, one with amplitude $+\frac{1}{2}$ and the other with amplitude $-\frac{1}{2}$. The probability of that outcome is not $\frac{1}{4} + \frac{1}{4} = \frac{1}{2}$. It is

$$\left|+\tfrac{1}{2} - \tfrac{1}{2}\right|^2 = 0$$

The outcome never happens. Two ways of reaching it have annihilated each other. This has no analogue anywhere in classical probability theory, and it is the entire physical resource that quantum computing exploits. Every quantum algorithm in existence is, at bottom, an arrangement of amplitudes such that the wrong answers cancel.

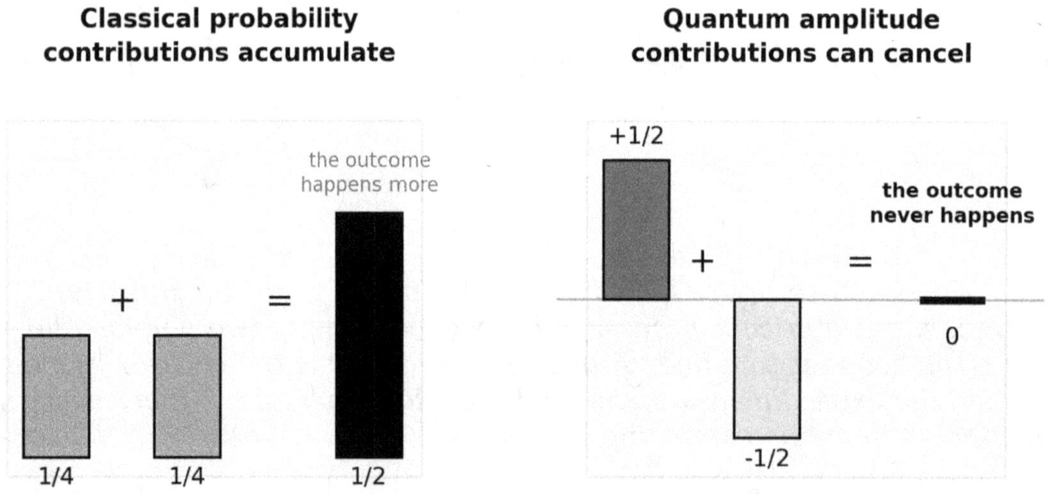


***Figure 2. Why amplitudes are not probabilities.*** *Two paths to the same outcome. Classically, two chances of one-quarter make one chance of one-half — probabilities only ever add. Quantum mechanically, amplitudes of $+\frac{1}{2}$ and $-\frac{1}{2}$ sum to zero, and the outcome is eliminated entirely. This cancellation has no counterpart in any classical theory, and it is the whole physical resource that quantum algorithms exploit.*

There is a striking structural argument for why nature uses the 2-norm — that is, why probability is the *squared* magnitude of the amplitude. Scott Aaronson poses the question as: for which $p$ does there exist an interesting family of linear transformations preserving the $p$-norm of a vector? The answer is that there are exactly two. For $p = 1$ the norm-preserving maps are the stochastic matrices, and one recovers classical probability theory. For $p = 2$ they are the unitary matrices, and one recovers quantum mechanics. For every other value of $p$, the constraints are so rigid that only permutations and sign flips survive — nothing rich enough to compute with. **[V]**[1]

In this sense quantum mechanics is not an arbitrary quirk of nature. It is the only other consistent linear theory of probability that exists.

### 2.3 Why complex, and not merely negative

Negative amplitudes would suffice for cancellation, so why does nature use complex ones? Two arguments are robust:

**Continuity of dynamics.** Physical evolution is continuous in time, which requires that every allowed transformation have a square root inside the group of allowed transformations — one must be able to apply "half" of any operation. Among real orthogonal $2 \times 2$ matrices, a reflection has no real square root. Extending to complex amplitudes closes this gap.

**Composition.** With complex amplitudes, the number of parameters describing a composite system multiplies correctly: a system of dimension $N_a$ combined with one of dimension $N_b$ has parameter count $(N_a N_b)^2$. Real-amplitude quantum mechanics fails this consistency check.

A third, empirical argument was advanced in 2021 by Renou and collaborators, who showed that real-amplitude quantum theory and standard complex quantum theory make *different* predictions in a network with two independent entangled sources, yielding a Bell-type inequality that real quantum theory cannot violate as strongly. **[V]**[2] This result is widely reported as having falsified real-valued quantum mechanics.

*A caution.* Craig Gidney has argued that the result depends critically on an assumption stated only in the supplementary material — that the parties share no prior entanglement — and that a real-amplitude model with pre-distributed entangled states reproduces the observed statistics. **[R]**[3] Readers should treat this as "real quantum theory is falsifiable under an independent-sources assumption" rather than as an unconditional result. The two structural arguments above are not subject to this caveat.

[1] S. Aaronson, *Quantum Computing Since Democritus*, Lecture 9. https://www.scottaaronson.com/democritus/lec9.html

[2] M.-O. Renou et al., "Quantum theory based on real numbers can be experimentally falsified," *Nature* 600, 625–629 (2021). https://www.nature.com/articles/s41586-021-04160-4

[3] C. Gidney, "Real quantum mechanics is not falsified." https://algassert.com/post/2501

## 3. Superposition, precisely

This section is the heart of Part I, because superposition is the concept most damaged by popular description.

### 3.1 What it means

A qubit is in superposition, **with respect to a chosen measurement**, when its state vector is not an eigenvector of the observable being measured. That is the complete technical content.

Note what this definition does not say. It does not say the qubit lacks a definite state. The state $|\psi\rangle = \alpha|0\rangle + \beta|1\rangle$ is a complete, maximally specified description of the qubit's physical condition. There is nothing further to know about it. What the qubit lacks is a definite *value of the particular observable* $Z$, and that is a different thing entirely.

Note also that superposition is basis-relative. The state $|0\rangle$ — the most definite, least mysterious state a qubit can be in — is itself an equal superposition in the $X$ basis:

$$|0\rangle = \frac{1}{\sqrt{2}}(|+\rangle + |-\rangle)$$

This observation alone disposes of the "both values at once" framing. If superposition meant that a system holds two values simultaneously, then every state of every system would be in superposition with respect to some basis, and the concept would carry no information.

### 3.2 What it does not mean

**It does not mean the qubit is secretly 0 or 1 and we simply don't know which.** That is a genuinely different physical state — a statistical mixture — and the two are experimentally distinguishable in a single round of measurement. A qubit in the state $|+\rangle = (|0\rangle + |1\rangle)/\sqrt{2}$ measured in the $X$ basis returns $+1$ with certainty. A qubit that is 0-or-1 with equal probability, measured in the same basis, returns $\pm 1$ at random. The formal distinction is captured in the density-matrix picture (§3.4): $|+\rangle$ sits on the surface of the Bloch ball; the coin-flip mixture sits at its centre.

**It does not mean the qubit "is 0 and 1 at the same time."** There is no sense in which two definite values are simultaneously instantiated. This is not merely a matter of interpretive taste; it is constrained by theorem. The Kochen–Specker theorem establishes that in Hilbert spaces of dimension three or higher, it is impossible to assign definite pre-existing values to all observables in a way that is independent of what else is measured alongside them. **[V]**[4] The original 1967 construction required 117 projectors; Cabello and collaborators later reduced this to 18 in dimension four. Crucially, this result requires no assumption about locality at all — it is a stronger constraint, in that respect, than Bell's theorem.

[4] S. Kochen and E. P. Specker, "The Problem of Hidden Variables in Quantum Mechanics," *J. Math. Mech.* 17, 59–87 (1967).

**It does not mean $n$ qubits explore $2^n$ possibilities in parallel.** This is the subject of Part II.

### 3.3 Relative phase is where the computation lives

Consider two states:

$$|+\rangle = \tfrac{1}{\sqrt{2}}(|0\rangle + |1\rangle) \qquad |-\rangle = \tfrac{1}{\sqrt{2}}(|0\rangle - |1\rangle)$$

They differ only in a minus sign — a relative phase. And yet they are orthogonal, meaning they can be distinguished perfectly, with certainty, in a single measurement. If you measure in the $X$ basis, $|+\rangle$ always gives one answer and $|-\rangle$ always gives the other.

Meanwhile, if you multiply an entire state by a phase factor, $e^{i\gamma}|\psi\rangle$, you get a state that is physically identical to $|\psi\rangle$ and cannot be distinguished from it by any measurement whatsoever. This is global phase, and it is not physical.

The distinction matters enormously. Popular accounts of superposition describe the magnitudes $|\alpha|$ and $|\beta|$ — "it's 70% zero and 30% one" — and omit the relative phase entirely. But the magnitudes are the classical-looking part. The relative phase is the genuinely quantum degree of freedom, and it is the one that does all the computational work. Every interference effect in Deutsch-Jozsa, Grover, and Shor is an arrangement of relative phases such that unwanted amplitudes cancel.

> A superposition without phase information is just a probability distribution. If one sentence in this paper is worth keeping, it may be that one.

### 3.4 The Bloch sphere

The two-real-parameter state space of a qubit can be drawn. Writing

$$|\psi\rangle = \cos(\theta/2)|0\rangle + e^{i\varphi}\sin(\theta/2)|1\rangle$$

with $\theta \in [0, \pi]$ and $\varphi \in [0, 2\pi)$, every pure qubit state corresponds to a point on the surface of a unit sphere, at polar angle $\theta$ and azimuthal angle $\varphi$. This is the Bloch sphere.

| State | Bloch vector | Meaning |
|---|---|---|
| $\|0\rangle$ | $(0,0,1)$ | north pole |
| $\|1\rangle$ | $(0,0,-1)$ | south pole |
| $\|+\rangle$ | $(1,0,0)$ | $+x$ |
| $\|-\rangle$ | $(-1,0,0)$ | $-x$ |
| $\|+i\rangle$ | $(0,1,0)$ | $+y$ |
| $\|-i\rangle$ | $(0,-1,0)$ | $-y$ |

Each axis corresponds to a measurement one could perform. The $z$-axis poles are the outcomes of a computational-basis ($Z$) measurement; the $x$-axis poles are the outcomes of an $X$ measurement; and so on. The components of the Bloch vector are precisely the expectation values of the three Pauli observables: $r_i = \langle \sigma_i \rangle$.

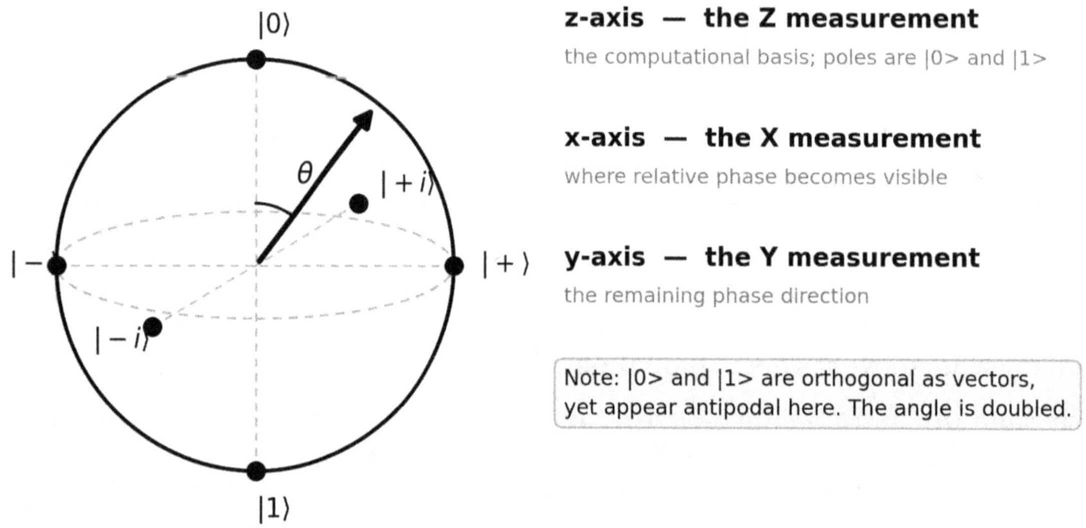


***Figure 3. The Bloch sphere: each axis is a question you could ask.*** *Every pure state of a single qubit is a point on this surface. The three axes correspond to three different measurements one could perform, and the Bloch vector's components are exactly the expected values of the three Pauli observables. The picture does not generalise: there is no faithful sphere for two or more qubits.*

Two subtleties are worth flagging, because both are routinely elided:

**The factor of two.** States that are orthogonal in Hilbert space — $|0\rangle$ and $|1\rangle$, which are at 90° to each other as vectors — appear *antipodal* on the Bloch sphere, 180° apart. The Bloch sphere is not the Hilbert space; it is a quotient of it, and the angle is doubled in the mapping.

**It does not generalise.** There is no faithful Bloch-sphere picture for two or more qubits. A two-qubit pure state has six real parameters, and no three-dimensional sphere captures it. The Bloch sphere is a single-qubit convenience, and intuitions built on it break down at exactly the point where quantum computing becomes interesting.

### 3.5 Two continuous parameters in, one bit out

Here is the fact that quietly governs everything in Part II.

A qubit's state is specified by two continuous real parameters, $\theta$ and $\varphi$. These are real, physical, and can in principle take any of a continuum of values. One might imagine encoding an enormous amount of information in them — the digits of $\pi$, say, in the binary expansion of $\theta$.

You cannot retrieve it. Measuring the qubit in any fixed basis returns exactly one classical bit. Three independent barriers enforce this:

1. **Non-orthogonal states cannot be perfectly distinguished.** Two states are reliably distinguishable if and only if they are orthogonal, and in a two-dimensional space there are at most two mutually orthogonal states.
2. **You cannot make copies to average over.** This is the no-cloning theorem (§4.3).
3. **Holevo's bound caps the total.** $n$ qubits can convey at most $n$ classical bits (§4.4).

Determining $\theta$ and $\varphi$ to precision $\varepsilon$ requires on the order of $1/\varepsilon^2$ identically prepared copies. The continuous parameters are physically real and operationally inaccessible in any single copy. This is the precise sense in which the "hidden" information inside a qubit is not information at all.

## 4. Measurement, and what cannot be extracted

### 4.1 The Born rule

When a qubit in state $\alpha|0\rangle+\beta|1\rangle$ is measured in the computational basis, the outcome is 0 with probability $|\alpha|^2$ and 1 with probability $|\beta|^2$.

This is the Born rule, introduced by Max Born in 1926. **[V]**[5] A charming detail of its origin: the main text of Born's paper states that the amplitude gives the probability; a footnote added in proof corrects this to the square of its magnitude. He received the Nobel Prize for it in 1954.

Is the rule derivable, or must it be assumed? Partially derivable. Gleason's theorem (1957) shows that any consistent, non-contextual assignment of probabilities to measurement outcomes in a Hilbert space of dimension three or greater must have the Born form. **[V]**[6] However — and this is often omitted — Gleason's theorem fails in dimension two, which is to say it does not apply to a single qubit. The gap was closed by Busch and independently by Caves, Fuchs, Manne and Renes, who extended the argument to generalised measurements. **[V]**[7]

### 4.2 What measurement does not explain

It is worth being honest about the limits here, because this is where physics genuinely runs out.

Decoherence (§6) explains why we never observe superpositions of macroscopic pointer readings. It does not explain why there is a single definite outcome at all.

[5] M. Born, "Zur Quantenmechanik der Stoßvorgänge," *Zeitschrift für Physik* 37, 863–867 (1926).

[6] A. M. Gleason, "Measures on the closed subspaces of a Hilbert space," *Indiana Univ. Math. J.* 6, 885–893 (1957).

[7] P. Busch, "Quantum States and Generalized Observables: A Simple Proof of Gleason's Theorem," *Phys. Rev. Lett.* 91, 120403 (2003).

The *Stanford Encyclopedia of Philosophy* is explicit on this point: decoherence explains why superpositions of measurement results are not observed, but not why any measurement result is observed in the first place. **[V]**[8]

This unresolved residue is the measurement problem, and it is precisely the gap into which the competing interpretations of quantum mechanics are inserted. We return to it in Part IV. For now the honest statement is: the Born rule is an axiom of the operational formalism, its form is derivable from weaker assumptions, and the question of why any particular outcome occurs remains open.

### 4.3 No-cloning

**Theorem.** There is no unitary operation that takes an arbitrary unknown quantum state $|\psi\rangle$ and a blank register to two copies of $|\psi\rangle$. **[V]**

The proof by linearity is instructive because it shows what you get instead of a copy. Suppose a machine copies the basis states:

$$U|0\rangle|s\rangle = |00\rangle \qquad U|1\rangle|s\rangle = |11\rangle$$

Linearity then forces its action on a superposition:

$$U\left(\alpha|0\rangle + \beta|1\rangle\right)|s\rangle = \alpha|00\rangle + \beta|11\rangle$$

But a genuine copy would be $(\alpha|0\rangle+\beta|1\rangle)\otimes(\alpha|0\rangle+\beta|1\rangle)$, which expands to include cross terms $\alpha\beta(|01\rangle + |10\rangle)$. Those terms are missing. What the machine has produced is not two copies of the state — it is an entangled pair. This is not a failure mode; it is a preview of §5.

The theorem was proved by Wootters and Zurek and independently by Dieks in 1982, in response to a proposal by Nick Herbert to use cloning for faster-than-light signalling. **[V]**[9] A complete proof had in fact been published by James Park in 1970 and overlooked for decades — a priority documented by Ortigoso in 2018.

No-cloning does not forbid everything. States drawn from a known orthogonal set can be copied freely — which is exactly why classical information can be copied. Imperfect cloning is possible: Bužek and Hillery showed a universal cloner achieving fidelity 5/6. And quantum error correction (§15) evades the theorem entirely by spreading information across an entangled system without ever copying it.

### 4.4 Holevo's bound

**Theorem (Holevo, 1973).** $n$ qubits can be used to transmit at most $n$ classical bits of retrievable information. **[V]**[10]

[8] G. Bacciagaluppi, "The Role of Decoherence in Quantum Mechanics," *Stanford Encyclopedia of Philosophy*. https://plato.stanford.edu/entries/qm-decoherence/

[9] W. K. Wootters and W. H. Zurek, "A single quantum cannot be cloned," *Nature* 299, 802–803 (1982).

[10] A. S. Holevo, "Bounds for the quantity of information transmitted by a quantum communication channel," *Problems of Information Transmission* 9, 177–183 (1973).

This deserves to be stated loudly, because it is the sharpest possible rebuttal to the idea that a quantum register's exponentially large state space constitutes exponentially large storage. A 300-qubit register is described by $2^{300}$ complex amplitudes — more numbers than there are atoms in the observable universe. It can deliver 300 bits.

Two qualifications:

**Superdense coding is not a counterexample.** Bennett and Wiesner showed that Alice can send two classical bits by transmitting one qubit — but only if she and Bob already share an entangled pair. Holevo's bound applies to the transmitted qubit alone; the total quantum resource consumed is two qubits. Without the pre-shared entanglement the protocol is impossible, precisely because it would violate the bound.

**Random access codes do no better.** Nayak's bound shows that even in the weaker setting where Bob need only recover any one bit of Alice's message on demand, quantum encoding offers no asymptotic advantage over classical.

## 5. Entanglement

### 5.1 The definition

A two-qubit state is entangled when it cannot be written as a product of individual qubit states. The canonical examples are the four Bell states:

$$|\Phi^{\pm}\rangle = \frac{1}{\sqrt{2}}(|00\rangle \pm |11\rangle) \qquad |\Psi^{\pm}\rangle = \frac{1}{\sqrt{2}}(|01\rangle \pm |10\rangle)$$

Each is maximally entangled, and each exhibits a property with no classical counterpart whatsoever: the joint state is completely specified while neither individual state is specified at all. If you compute the state of either qubit alone, you get the maximally mixed state — total ignorance. The composite system is in a pure state of maximal knowledge; its parts are in states of zero knowledge.

Classically this is impossible. Complete knowledge of a pair of bits entails complete knowledge of each bit. Entanglement is precisely the situation in which information resides in the correlations and nowhere else.

### 5.2 Why it is not just correlation

Two socks in separate boxes are correlated: open one, learn the other. This is not mysterious, and it is not entanglement. The distinction was made empirically testable by John Bell in 1964. **[V]**[11]

Bell's argument assumes only that (i) outcomes are determined by local properties carried by the particles, and (ii) the choice of what to measure is independent of those properties. From these, he derived an inequality constraining the correlations

[11] J. S. Bell, "On the Einstein Podolsky Rosen Paradox," *Physics* 1(3), 195–200 (1964).

any such theory can produce. In its most-used form, due to Clauser, Horne, Shimony and Holt: **[V]**[12]

| Bound on the CHSH quantity $S$ | Value |
|---|---|
| Local hidden-variable theories | $\lvert S\rvert \le 2$ |
| Quantum mechanics (Tsirelson) | $\lvert S\rvert \le 2\sqrt{2} \approx 2.828$ |
| Mere no-signalling | $\lvert S\rvert \le 4$ |

Nature violates the classical bound and saturates the quantum one. Why it stops at $2\sqrt{2}$ rather than 4 is an open foundational question: Popescu and Rohrlich constructed hypothetical "PR boxes" that reach 4 while still forbidding faster-than-light signalling, showing that no-signalling alone does not single out quantum mechanics.

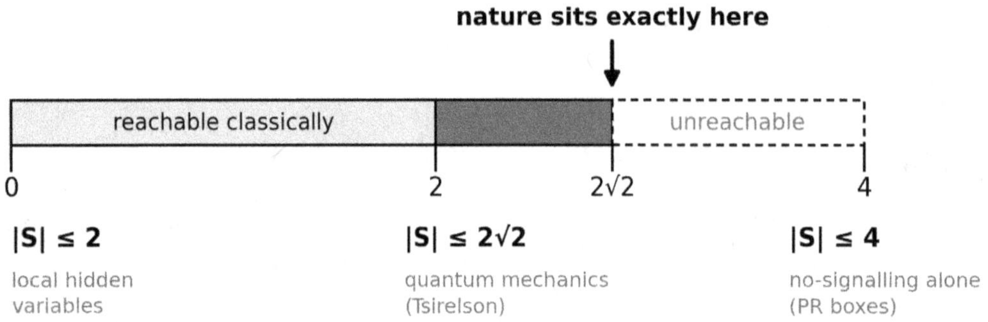


***Figure 4. Three bounds, and where nature actually sits.*** *Local hidden-variable theories cannot exceed 2. Quantum mechanics cannot exceed $2\sqrt{2}$, and experiment finds nature saturating exactly that value. Nothing forbids correlations up to 4 on no-signalling grounds alone — which means quantum mechanics is more constrained than relativity requires, and why it stops where it does remains an open foundational question.*

**The loopholes are closed.** Three experiments in 2015 — Hensen and colleagues in Delft using nitrogen-vacancy centres 1.3 km apart, and photonic experiments by Giustina and by Shalm — simultaneously closed the detection, locality and freedom-of-choice loopholes. **[V]**[13] The photonic results reached extraordinary significance ($p \approx 10^{-31}$ in the Vienna experiment). A 2018 "cosmic Bell test" went further, deriving measurement settings from the light of quasars at redshifts up to 3.911 — photons emitted over 12 billion years ago — thereby excluding local-realist conspiracies from roughly 96% of the relevant spacetime volume. **[V]**[14]

[12] J. F. Clauser, M. A. Horne, A. Shimony and R. A. Holt, *Phys. Rev. Lett.* 23(15), 880–884 (1969).

[13] B. Hensen et al., *Nature* 526, 682–686 (2015); M. Giustina et al., *Phys. Rev. Lett.* 115, 250401 (2015); L. K. Shalm et al., *Phys. Rev. Lett.* 115, 250402 (2015).

[14] D. Rauch et al., "Cosmic Bell Test Using Random Measurement Settings from High-Redshift Quasars," *Phys. Rev. Lett.* 121, 080403 (2018).

The 2022 Nobel Prize in Physics went to Alain Aspect, John Clauser and Anton Zeilinger for this body of work.

**What Bell's theorem does not show**, since this is widely misreported: it does not permit superluminal signalling (the marginal statistics at each end are unchanged by the distant setting); it does not rule out hidden variables in general (Bohmian mechanics is an explicitly non-local hidden-variable theory in good standing); and it does not tell you which of its assumptions to abandon.

### 5.3 Monogamy

Entanglement cannot be freely shared. If two qubits are maximally entangled with each other, neither can be entangled with anything else at all — a constraint made quantitative by Coffman, Kundu and Wootters and generalised to $n$ qubits by Osborne and Verstraete. **[V]**[15]

This has no classical analogue: a classical random bit can be broadcast to arbitrarily many parties, all perfectly correlated. Monogamy is the structural reason device-independent quantum key distribution can be secure — an eavesdropper's correlation with the key is bounded by the legitimate parties' correlation with each other.

## 6. Decoherence: why this is hard

Decoherence is not an exotic additional process. It is ordinary quantum mechanics applied honestly to a system that is not isolated.

Take a qubit in superposition, together with an environment in some initial state:

$$(\alpha|0\rangle + \beta|1\rangle) \otimes |E_0\rangle$$

Any interaction that depends on the qubit's state will correlate the two:

$$\alpha|0\rangle|E_0\rangle + \beta|1\rangle|E_1\rangle$$

The environment has now recorded which state the qubit is in. If we ask what the qubit alone looks like — the environment being unobserved — the off-diagonal terms of its density matrix are multiplied by the overlap $\langle E_1|E_0\rangle$. As the environment's two records become distinguishable, that overlap goes to zero, and the qubit's description degrades from a coherent superposition to a classical probabilistic mixture. On the Bloch sphere, the state vector shrinks from the surface toward the centre.

Nothing collapsed. Nothing was measured, in the sense of a conscious observer looking. The phase information did not vanish; it leaked into correlations with the environment, where it is no longer accessible to us.

[15] V. Coffman, J. Kundu and W. K. Wootters, "Distributed entanglement," *Phys. Rev. A* 61, 052306 (2000); T. J. Osborne and F. Verstraete, *Phys. Rev. Lett.* 96, 220503 (2006).

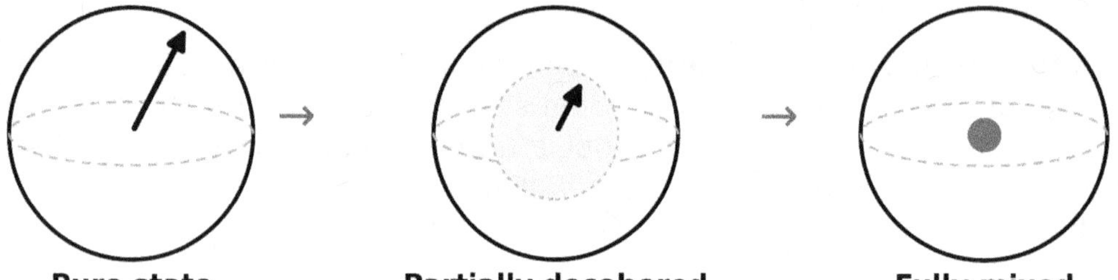


***Figure 5. Decoherence: nothing collapses, the phase simply leaks away.*** *Environmental interaction correlates the qubit with degrees of freedom we cannot track. The phase information is not destroyed — it becomes distributed in correlations with the environment, where it is no longer accessible. Because the rate grows with the number of environmental degrees of freedom, the problem becomes harder as machines get larger, which is precisely the wrong direction for an engineer.*

Two timescales characterise this in real hardware:

$T_1$ — **energy relaxation.** The excited state decays to the ground state, dumping energy into the environment.

$T_2$ — **dephasing.** The relative phase between $|0\rangle$ and $|1\rangle$ randomises.

These are related by $1/T_2 = 1/(2T_1) + 1/T_\varphi$, where $T_\varphi$ captures pure dephasing — phase noise with no energy exchange. An immediate corollary: $T_2 \leq 2T_1$. Energy loss necessarily destroys phase, though the converse does not hold.

The engineering significance is hard to overstate. Zurek's analysis shows that decoherence rates grow with the number of environmental degrees of freedom, which means the problem gets worse as systems get larger — precisely the opposite of what an engineer building a bigger machine would like. **[V]**[16] His illustrative calculation for Hyperion, a chaotically tumbling moon of Saturn, gives a coherence breakdown time of roughly twenty years, which is a vivid way of seeing why the macroscopic world looks classical.

## 7. The exponential state space, stated carefully

An $n$-qubit register is described by $2^n$ complex amplitudes:

$$|\psi\rangle = \sum_{x \in \{0,1\}^n} \alpha_x |x\rangle$$

The growth is brutal. Storing the state vector in double precision requires $2^{(n+4)}$ bytes:

[16] W. H. Zurek, "Decoherence, einselection, and the quantum origins of the classical," *Rev. Mod. Phys.* 75, 715–775 (2003).

| Qubits | Memory to store the state vector |
|---|---|
| 20 | 16 MiB |
| 30 | 16 GiB — a laptop |
| 40 | 16 TiB |
| 50 | 16 PiB |
| 60 | 16 EiB |
| 300 | ~$10^{79}$ bytes — more than the atoms in the observable universe |

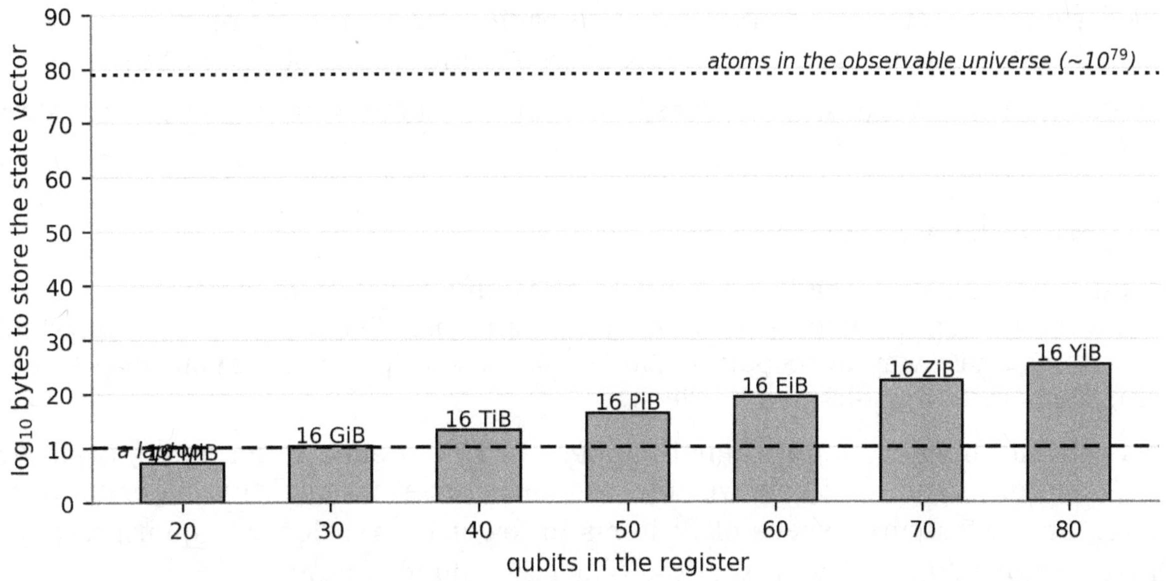


***Figure 6. The description grows exponentially. The computational power does not.*** *Storing the state vector of an n-qubit register takes* $2^{(n+4)}$ *bytes, which passes a laptop's memory around 30 qubits and the atom count of the observable universe well before 300. This table is entirely correct as a statement about the description of the state, and it licenses almost nothing about what the machine can compute.*

This table is the source of nearly all popular intuition about quantum computing, and nearly all of the resulting confusion. It is entirely correct as a statement about the *description* of the state. It licenses almost nothing about computational power. Part II explains why.

---

# Part II — How the Speedup Actually Works

## 8. The misconception, stated plainly

Here is the standard popular account:

> A classical computer must try each possible solution one at a time. A quantum computer places its qubits in a superposition of all $2^n$ possible answers, evaluates them all simultaneously in parallel, and returns the correct one. It is as though $2^n$ computers were running at once.

This is wrong. Not oversimplified — wrong, in a way that inverts what the machine actually does and licenses conclusions that are provably false.

Scott Aaronson has spent two decades correcting it; his blog carries a standing note to the effect that quantum computers do not solve hard problems by trying all solutions in parallel. His positive statement of the mechanism is the clearest available: with every quantum algorithm, the aim is to choreograph a pattern of interference in which the contributions to each wrong answer's amplitude cancel out while the contributions to the right answer's amplitude reinforce. **[V]**[17]

The distinction is not aesthetic. Consider what would follow if the parallelism story were true.

### 8.1 The argument that settles it

If a quantum computer really did evaluate $2^n$ candidate solutions in parallel, then unstructured search would be essentially instantaneous. Give the machine a haystack of $2^n$ items and a way to recognise the needle; every "parallel branch" checks one item; the answer appears in a single step.

This does not happen, and it cannot happen. In 1997 Bennett, Bernstein, Brassard and Vazirani proved that relative to a random oracle, no quantum algorithm can search an unstructured space of $N$ items in fewer than order $\sqrt{N}$ queries. **[V]**[18] Grover's algorithm achieves exactly $\sqrt{N}$, so the bound is tight.

Square root. Not logarithm, not constant — square root. To search a space of a trillion items, a quantum computer needs about a million steps, not one.

The BBBV authors anticipated the misconception in the paper itself. They note that it is tempting to try to use the apparent exponential parallelism to simulate nondeterminism, and that the formalism of quantum mechanics imposes inherent constraints on the scope of that parallelism. They then show exactly where the naive construction fails: a direct quantum analogue of the classical algorithm would begin in an equally weighted superposition of $2^n$ computation paths, query a different string on each path, and finally collapse the superposition by asking whether the query had found the nonempty location — which yields a success probability of $1/2^n$, precisely what a classical computer achieves.

There it is. You can put the register into a superposition of all $2^n$ candidates. You can evaluate the function on all of them. And when you measure, you get a uniformly

[17] S. Aaronson, "Quantum Computing: Between Hope and Hype," *Shtetl-Optimized*, 22 September 2024. https://scottaaronson.blog/?p=8329

[18] C. H. Bennett, E. Bernstein, G. Brassard and U. Vazirani, "Strengths and Weaknesses of Quantum Computing," *SIAM J. Comput.* 26(5), 1510–1523 (1997). https://arxiv.org/abs/quant-ph/9701001

random one — exactly as useless as guessing. The superposition was created and destroyed with no benefit whatsoever.

The parallelism is real. Harvesting it is the problem, and superposition alone does not solve it.

### 8.2 Why linearity forbids the shortcut

The deepest reason the parallel-universe picture cannot be cashed in was stated by Aaronson in *Scientific American*: the Schrödinger equation is linear, and if one thinks of the paths in a superposition as parallel universes, it is exactly that linearity which prevents the universe that happened to find the answer from telling all the others about it. **[V]**[19]

Quantum evolution is linear. There is no operation available that inspects the branches, notices which one succeeded, and amplifies it. Such an operation would be nonlinear, and nonlinear quantum mechanics is not the physics we have. It is worth noting that if it were, quantum computers would solve NP-complete problems trivially — Abrams and Lloyd showed exactly this in 1998, which is a good reason to be sceptical of nonlinear modifications.

Aaronson's image for the misconception is worth keeping: it is as though the computer conjures an exponentially powerful monster into existence, which then scurries into a closet the moment the lights come on.

## 9. What actually happens: interference

The correct picture is this. A quantum algorithm is a carefully designed sequence of operations that:

1. Creates a superposition over many computational paths.
2. Arranges, through the phases accumulated along those paths, that amplitudes leading to wrong answers destructively interfere and cancel.
3. Arranges that amplitudes leading to right answers constructively interfere and add.
4. Measures, now finding the right answer with high probability.

The design difficulty — the thing that makes quantum algorithms rare and hard to invent — is steps 2 and 3 together, performed *without knowing the answer in advance*. You must construct an interference pattern that concentrates amplitude on a solution you cannot identify. This is only possible when the problem has mathematical structure that can be exploited blindly.

[19] S. Aaronson, "The Limits of Quantum Computers," *Scientific American*, March 2008. Author draft: https://www.scottaaronson.com/writings/limitsqc-draft.pdf

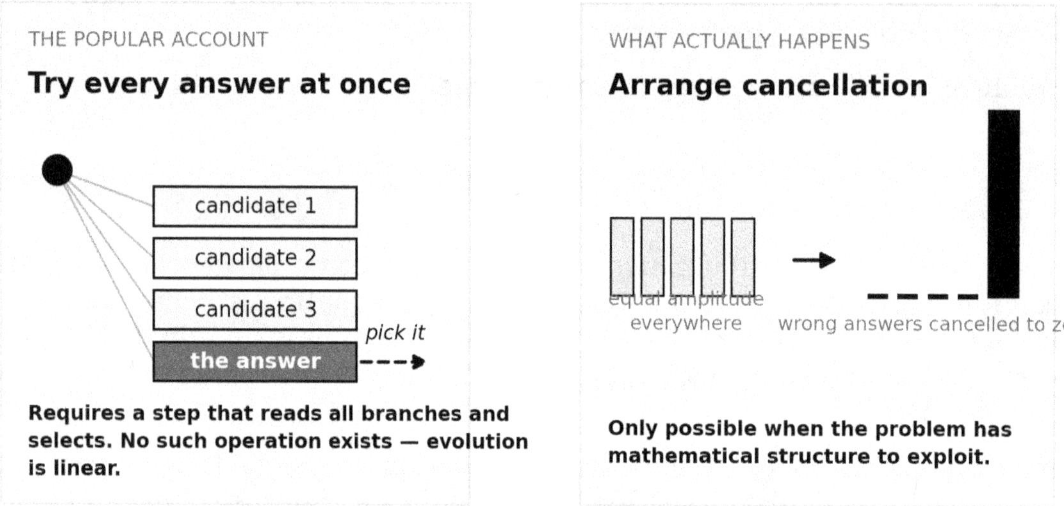


***Figure 7. Two accounts of the same machine. Only one of them is true.*** *The left panel is the account found in most popular coverage. It requires a nonlinear "inspect and select" operation that quantum mechanics does not provide. The right panel is the mechanism: interference sculpts the probability distribution before measurement, so that a single sample is likely to be the answer. The difference is not one of emphasis — it is the difference between a machine that would break NP-complete problems and the machine we actually have.*

Aaronson calls this the **Law of Conservation of Weirdness**: every problem admitting exponential quantum speedup must have some unusual structure permitting amplitude to be concentrated on rare correct answers. **[V]**[20] Generic search has no such structure, and gets no such speedup.

## 10. Grover's algorithm: why the square root

Grover's algorithm is worth working through because its mechanism makes the whole point quantitative. **[V]**[21]

Start with all $N$ items in equal superposition; each has amplitude $1/\sqrt{N}$. Each iteration of the algorithm does two things: it flips the sign of the marked item's amplitude, then it reflects all amplitudes about their mean. Geometrically, in the two-dimensional plane spanned by "the solution" and "everything else," each iteration **rotates the state vector by a fixed small angle**.

That fixed angle is the entire story. The solution's amplitude grows by roughly $2/\sqrt{N}$ per iteration — a constant increment, not a doubling. To bring the amplitude from $1/\sqrt{N}$ up to something near 1, you need about $(\pi/4)\sqrt{N}$ iterations. The amplitude has to be walked out of the hole one step at a time.

[20] S. Aaronson, "How Much Structure Is Needed for Huge Quantum Speedups?", Solvay Conference (2022). https://www.scottaaronson.com/papers/aarsolvay.pdf

[21] L. K. Grover, "A fast quantum mechanical algorithm for database search," STOC (1996). https://arxiv.org/abs/quant-ph/9605043

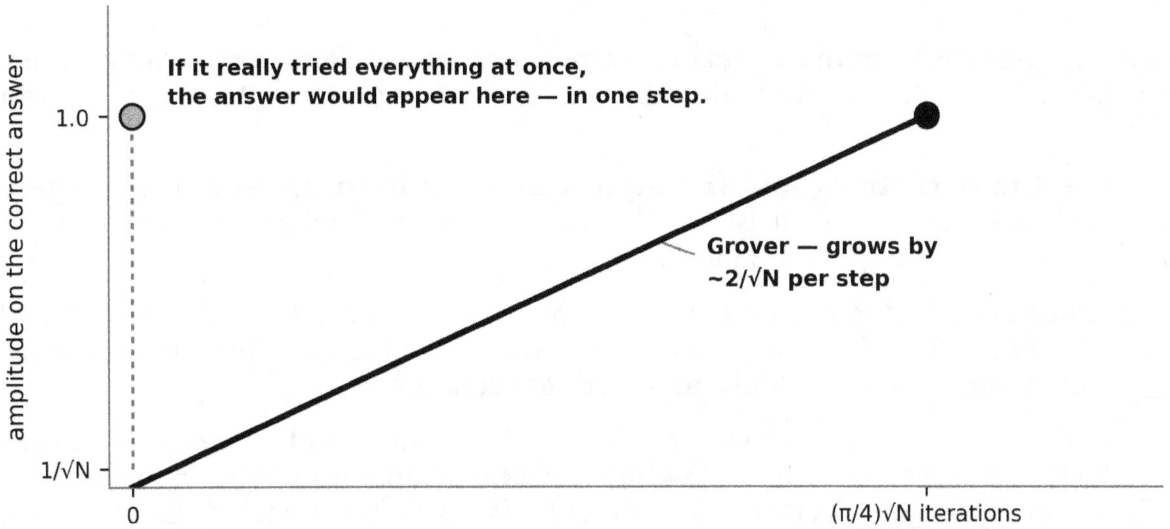


***Figure 8. Grover's algorithm: the amplitude has to be walked out, one step at a time.*** *Each iteration rotates the state by a fixed small angle, so the amplitude on the correct answer grows by a roughly constant increment — not by doubling. Climbing from* $1/\sqrt{N}$ *to near 1 therefore takes* $(\pi/4)\sqrt{N}$ *steps. A machine that genuinely evaluated all candidates in parallel would need no climb at all.*

A machine that genuinely tried everything at once would require no such walk. Grover's algorithm needs $\sqrt{N}$ steps precisely *because* it is not doing that.

And this is not a limitation of Grover's cleverness. The BBBV bound proves $\sqrt{N}$ is optimal; Zalka later showed Grover's algorithm is optimal including the constant factor. **[V]**[22] Aaronson's summary is memorable: **Grover's algorithm was proven optimal before it was discovered to exist**, since the BBBV result circulated in 1994 and Grover published in 1996.

## 11. Shor's algorithm: structure, not brute force

Shor's algorithm factors integers in polynomial time, and it is the reason anyone outside physics pays attention to quantum computing. **[V]**[23] It does not work by trying all the factors.

The algorithm reduces factoring to a different problem: finding the **period** of the function $f(x) = a^x \bmod N$. Periodicity is a global algebraic property of that function, and the quantum Fourier transform is exquisitely suited to detecting it. When the QFT is applied, amplitudes at frequencies that do not match the period cancel by destructive interference; amplitudes at multiples of the true frequency reinforce. A single measurement then samples from a distribution sharply peaked at the answer, and classical post-processing (continued fractions) recovers the period.

[22] C. Zalka, "Grover's quantum searching algorithm is optimal," *Phys. Rev. A* 60, 2746 (1999).

[23] P. W. Shor, "Polynomial-Time Algorithms for Prime Factorization and Discrete Logarithms on a Quantum Computer," *SIAM J. Comput.* 26(5), 1484–1509 (1997).

Three observations matter for our argument:

**The exponentially many function values are never read out.** They are consumed in shaping a probability distribution. You get one sample from that distribution.

**Remove the periodicity and the algorithm does nothing.** It is not a general-purpose factoring engine; it is a period-finding engine applied to a problem that happens to reduce to period-finding.

**Factoring is not NP-complete.** It sits in NP ∩ coNP, and is not believed to be NP-hard. Breaking RSA does not imply solving the travelling salesman problem. This is a distinction the popular literature routinely collapses.

The generalisation is instructive. Shor's algorithm, Simon's algorithm and discrete logarithm are all instances of the **Abelian Hidden Subgroup Problem**, which quantum computers solve efficiently. The *non-abelian* case has resisted thirty years of effort — the dihedral case would break lattice-based post-quantum cryptography, and the best known result is Kuperberg's subexponential (not polynomial) algorithm. **[V]**[24]

If quantum computers simply tried everything at once, this sharp abelian/non-abelian boundary would be inexplicable. The speedup tracks a specific algebraic structure, not a general capability.

## 12. The theorems that bound the whole enterprise

Several results together establish that the restriction to structured problems is not a temporary state of ignorance but a proven feature.

**No superpolynomial speedup for total functions.** For any Boolean function defined on all inputs — no promise, no guaranteed structure — the classical and quantum query complexities are polynomially related. Beals and colleagues proved $D(f) = O(Q(f)^6)$ in 2001; Aaronson, Ben-David, Kothari, Rao and Tal tightened this to the optimal $D(f) = O(Q(f)^4)$ in 2021. **[V]**[25] Exponential speedups exist only for *promise* problems — problems where the input is guaranteed to have special structure.

**Speedups plausibly require structure even against random oracles.** Aaronson and Ambainis conjectured that any quantum algorithm can be simulated classically on most inputs with polynomial overhead, which if true would mean one essentially cannot separate P from BQP relative to a random oracle. **[V]**[26] The conjecture remains open — a claimed 2019 proof was retracted after a flaw was found — but partial results continue to accumulate.

[24] G. Kuperberg, "A Subexponential-Time Quantum Algorithm for the Dihedral Hidden Subgroup Problem," *SIAM J. Comput.* 35(1), 170–188 (2005).

[25] R. Beals et al., "Quantum Lower Bounds by Polynomials," *JACM* 48(4), 778–797 (2001); S. Aaronson, S. Ben-David, R. Kothari, S. Rao and A. Tal, STOC (2021). https://arxiv.org/abs/2010.12629

[26] S. Aaronson and A. Ambainis, "The Need for Structure in Quantum Speedups," *Theory of Computing* 10, 133–166 (2014). https://www.scottaaronson.com/papers/struc.pdf

**Complexity containment.** BQP, the class of problems efficiently solvable by quantum computers, is contained in PSPACE. A quantum computer can therefore be simulated classically using only polynomial space (though exponential time). Quantum computers compute nothing that classical computers cannot; the claim is entirely about efficiency. Moreover — and this surprises people — BQP $\neq$ P has never been proven. The exponential speedup of Shor's algorithm over the best known classical factoring algorithm is conjectural, not proven, because factoring is not known to be classically hard.

**Whether quantum computers can solve NP-complete problems is formally open, but essentially nobody expects it.** The evidence is BBBV: relative to a random oracle, they cannot.

**Entanglement and huge state spaces are not sufficient.** This is the decisive counterexample. The Gottesman–Knill theorem shows that circuits built only from Hadamard, CNOT and phase gates — which generate enormous entanglement across a $2^n$-dimensional space — are efficiently simulable on a classical computer. **[V]**[27] Aaronson and Gottesman gave an algorithm running in $O(n^2)$ per measurement.

Read that again. You can have exponentially many "branches," massive entanglement, the full $2^n$ state space — and a laptop can simulate it. Whatever produces quantum advantage, it is demonstrably not the size of the state space or the presence of entanglement.

## 13. What quantum computers are actually good for

Given all of the above, the honest list is short. Aaronson's own enumeration names three areas: **[V]**[28]

1. **Simulation of quantum physics and chemistry** — Feynman's original 1982 motivation, and the most defensible application, since a quantum system is being used to model a quantum system. **[V]**[29]
2. **Breaking currently deployed public-key cryptography** — real, consequential, and narrow. Symmetric cryptography and hash functions are essentially unaffected.
3. **Modest benefits for some optimisation problems** — typically a Grover-type quadratic speedup, which after error-correction overhead may not be worth having.

[27] S. Aaronson and D. Gottesman, "Improved simulation of stabilizer circuits," *Phys. Rev. A* 70, 052328 (2004). https://www.scottaaronson.com/papers/chp6.pdf

[28] S. Aaronson, "More on whether useful quantum computing is 'imminent'," *Shtetl-Optimized*, 21 December 2025. https://scottaaronson.blog/?p=9425

[29] R. P. Feynman, "Simulating Physics with Computers," *Int. J. Theor. Phys.* 21(6/7), 467–488 (1982).

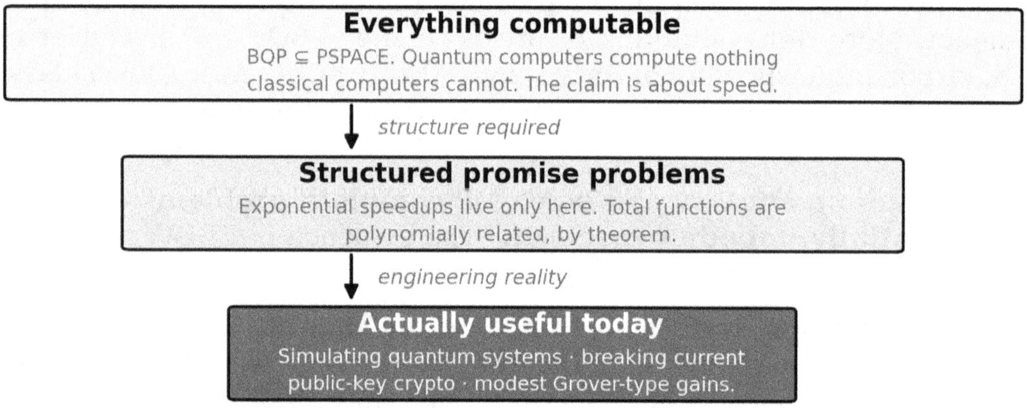


***Figure 9. The scope of the machine, drawn to scale.*** *Quantum computers compute nothing that classical computers cannot — BQP sits inside PSPACE, so the entire claim is about efficiency. Exponential speedups exist only for structured promise problems, by theorem. And what survives contact with error-correction overhead is narrower still.*

Two cautionary results deserve mention because they come from within the field.

**Even quantum chemistry may not deliver generic exponential advantage.** An eighteen-author study including John Preskill and Ryan Babbush — researchers with every professional incentive to claim otherwise — examined the evidence for exponential quantum advantage in ground-state energy estimation and concluded that evidence for such an advantage across chemical space has yet to be found, adding that while quantum computers may still prove useful for quantum chemistry, it may be prudent to assume exponential speedups are not generically available for this problem. **[V]**[30]

**Several claimed quantum machine-learning speedups have been "dequantized."** In 2018 Ewin Tang, then an undergraduate, showed that a flagship quantum recommendation-systems algorithm had a classical equivalent of comparable speed. **[V]**[31] The insight generalised: many quantum machine-learning speedups assumed the quantum algorithm could prepare states in logarithmic time, while denying the classical algorithm an analogous sampling access. Grant both sides the same input privileges and the exponential separation evaporates. This is a methodological lesson with teeth — though note that dequantization eats the low-rank linear-algebra speedups and does not touch Shor.

[30] S. Lee et al., "Evaluating the evidence for exponential quantum advantage in ground-state quantum chemistry," *Nature Communications* 14, 1952 (2023). https://arxiv.org/abs/2208.02199

[31] E. Tang, "A quantum-inspired classical algorithm for recommendation systems," STOC (2019). https://arxiv.org/abs/1807.04271

## Part III — Building One

### 14. Physical implementations

A qubit is any two-level quantum system that can be prepared, manipulated and measured while remaining isolated from its environment. Several physical realisations are being pursued in parallel; none has clearly won.

**Superconducting transmons** (Google, IBM, Rigetti) use a Josephson junction shunted by a capacitor to form an anharmonic oscillator, with $|0\rangle$ and $|1\rangle$ as its two lowest energy levels, operated at 10–20 millikelvin. Gates are microwave pulses lasting tens of nanoseconds. Google's Willow processor reported mean $T_1$ of 68 µs and $T_2$ of 89 µs across 105 qubits. **[V]**[32] A 2025 Princeton result achieved 1.68 ms $T_1$ on an individual tantalum-on-silicon transmon — roughly 25× better than the mean qubit in a working processor, which is itself the practical story of the modality. **[V]**[33] *Advantages:* lithographic fabrication, fast gates, the most mature path to large 2D arrays. *Limitations:* short coherence, nearest-neighbour connectivity, enormous cryogenic overhead.

**Trapped ions** (Quantinuum, IonQ) confine individual atomic ions in electromagnetic traps, encoding $|0\rangle$ and $|1\rangle$ in hyperfine ground states. Because both states sit in the electronic ground manifold, $T_1$ is effectively infinite. Quantinuum's Helios, published in *Nature* in June 2026 and independently benchmarked by Sandia National Laboratories under a long-standing cooperative agreement, reports 98 barium-137 qubits with average single-qubit infidelity $2.5 \times 10^{-5}$, two-qubit infidelity $7.9 \times 10^{-4}$, and all-to-all connectivity. **[V]**[34] *Advantages:* identical qubits by construction, the highest fidelities of any platform, hours-scale coherence. The limitation is speed — Helios averages roughly 55 ms per circuit layer against about 1 µs for a superconducting error-correction cycle, some four to five orders of magnitude slower.

**Neutral atoms** (QuEra, Pasqal, Atom Computing) hold laser-cooled atoms in optical tweezer arrays that can be physically rearranged mid-computation, giving effectively arbitrary connectivity. Arrays of thousands of atoms have been demonstrated, including continuous operation of a coherent 3,000-atom system. **[V]** Two-qubit gates use the Rydberg blockade mechanism. This is currently the cheapest route to large qubit counts, at the cost of a slow clock.

**Photonic** (PsiQuantum, Xanadu) encodes qubits in single photons, with the appeal of room-temperature optical paths and semiconductor-foundry manufacturing. The

[32]Google Quantum AI and Collaborators, "Quantum error correction below the surface code threshold," *Nature* 638, 920–926 (2025). DOI 10.1038/s41586-024-08449-y. https://www.nature.com/articles/s41586-024-08449-y — the source both for the Willow coherence figures in §14 and the below-threshold result in §16.

[33]W. Bland et al., "Millisecond lifetimes and coherence times in 2D transmon qubits," *Nature* 647, 343–348 (2025). https://www.nature.com/articles/s41586-025-09687-4

[34]A. Ransford et al., "A 98-qubit trapped-ion quantum computer with all-to-all connectivity," *Nature* 655, 81–86 (2026). DOI 10.1038/s41586-026-10676-4. Preprint: arXiv:2511.05465. Independent benchmarking by Sandia National Laboratories under CRADA, reported June 2026.

binding constraint is photon loss, which is multiplicative through every component and has no analogue of "just improve $T_1$." Xanadu's Aurora system demonstrated 12 qubits across 35 chips and 13 km of fibre. **[R]**

**Silicon spin qubits** (Intel, Diraq) encode information in the spin of a single electron in a quantum dot, with the enormous strategic advantage of reusing the existing CMOS industry. A 2025 *Nature* result demonstrated above-99% fidelities on a 300 mm foundry line. **[V]**[35] Scale remains at single digits — roughly where superconducting qubits were in 2016.

**Topological qubits** (Microsoft) would encode information non-locally in Majorana zero modes, making it intrinsically protected from local noise. This remains contested, and the dispute sharpened in 2026. Microsoft's February 2025 *Nature* paper stated that its own measurements did not by themselves determine whether the low-energy states detected by interferometry were topological, while the accompanying press release claimed confirmation — a mismatch that generated substantial controversy. **[V]** In June 2026 *Nature* published a critique by Henry Legg arguing that the topological gap protocol is not robust, together with a reply from Microsoft Quantum in the same issue. **[V]**[36] Microsoft announced a second-generation chip, Majorana 2, on 2 June 2026, citing a preprint reporting coherence times above twenty seconds and roughly a thousandfold reliability improvement over its predecessor. **[P]** As of August 2026 that work has not cleared peer review, no non-Microsoft group has replicated the topological claim, and a related 2018 paper claiming definitive Majorana evidence was retracted. **[V]**

**Assessment [A].** The modality race is not converging, and the two leading platforms are failing in opposite directions. Trapped ions have the fidelities and lack the clock speed; superconductors have the clock speed and lack the fidelities. Neutral atoms have the most attractive scaling story and the least mature error correction. Any roadmap that presents a single winner is selling something.

[35] W. H. Steinacker et al., "Industry-compatible silicon spin-qubit unit cells exceeding 99% fidelity," *Nature* 646 (2025). https://www.nature.com/articles/s41586-025-09531-9

[36] H. F. Legg, "On the robustness of topological gap detection via transport," *Nature* 654, E22–E26 (2026), with a reply from Microsoft Quantum in the same issue. Earlier: Microsoft Azure Quantum, "Interferometric single-shot parity measurement in InAs–Al hybrid devices," *Nature* 638, 651–655 (2025). Majorana 2 announced 2 June 2026; preprint not peer reviewed at time of writing.

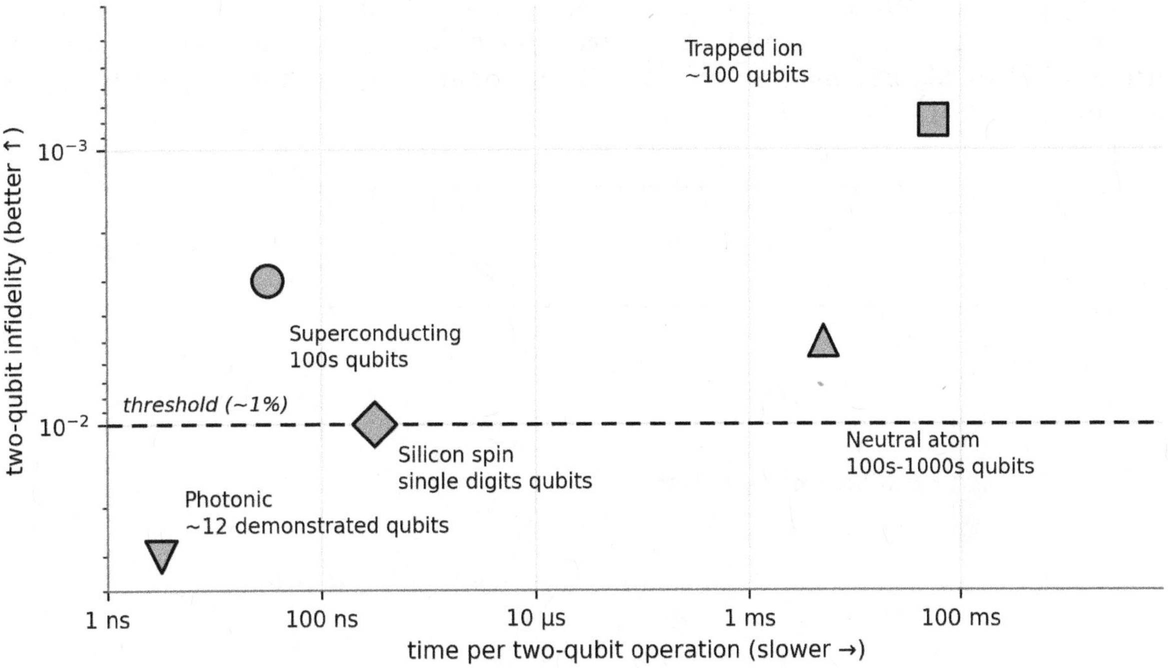


***Figure 10. The modality trade-off, drawn on the two axes that actually bind.*** *Trapped ions have the best two-qubit fidelities of any platform and a clock four to five orders of magnitude slower than superconducting hardware. Superconductors have the speed and not the fidelity. Nothing sits in the upper-left corner where a fault-tolerant machine would most like to be. Qubit counts are indicative of demonstrated systems, not of roadmap targets.*

## 15. Error correction, and what a "logical qubit" is

Decoherence would destroy any long computation. The solution is quantum error correction, which faces an immediate obstacle: you cannot copy quantum information (no-cloning, §4.3), and errors are continuous rather than discrete.

The resolution is elegant. Encode one logical qubit across many physical qubits, then repeatedly measure **parity checks** — operators that reveal which error occurred without revealing, and thereby collapsing, the encoded information. The measurement discretizes a continuous error into a discrete one, which can then be corrected.

The **threshold theorem** guarantees that if the physical error rate is below some constant, arbitrarily long computation is possible with only polylogarithmic overhead. **[V]**[37] For the surface code the threshold is approximately 1%, and the standard estimate is 1,000–10,000 physical qubits per logical qubit.

That overhead figure is the field's central engineering problem, and it has been moving fast. In 2019 Gidney and Ekerå estimated that factoring RSA-2048 would

[37] D. Aharonov and M. Ben-Or, *SIAM J. Comput.* 38(4), 1207–1282; E. Knill, R. Laflamme and W. H. Zurek, *Science* 279, 342–345 (1998); A. Yu. Kitaev, *Ann. Phys.* 303, 2–30 (2003).

need roughly 20 million noisy qubits. In May 2025 Gidney reduced that estimate to under one million qubits in under a week — a 95% reduction achieved entirely through better algorithms and codes, with no change in the underlying hardware assumptions. **[V]**[38]

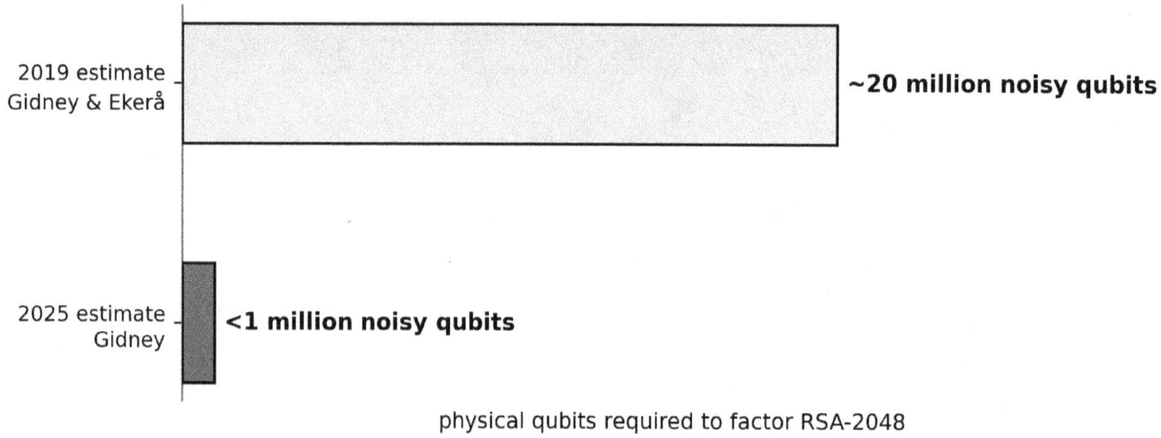


***Figure 11. The RSA-2048 resource estimate fell 95% without a single hardware improvement.*** *Both figures are estimates of physical qubits required, six years apart, under comparable hardware assumptions. The reduction came entirely from better algorithms and codes. This is the strongest available argument that hardware roadmaps are the wrong thing to watch alone — the software side of the overhead problem has been moving faster than the machines.*

**A warning about terminology.** "Logical qubit" is heavily abused in press releases. An error-*detecting* code that aborts when it notices a problem is not the same object as an error-*correcting* code that runs to completion, and neither is the same as a code demonstrated to suppress errors as its distance grows. Announcements should be read with that distinction in mind — and, as §16 shows, with attention to what fraction of the runs were thrown away.

## 16. Where the field stands, August 2026

The load-bearing results, distinguishing peer-reviewed work from vendor announcements:

**Google** demonstrated the first convincing below-threshold error correction in December 2024: a distance-7 surface code on Willow with logical error 0.143% per cycle, and — the critical number — an error suppression factor $\Lambda = 2.14$, meaning logical error halves each time the code grows. **[V]**[39] That is the signature of being

[38] C. Gidney, "How to factor 2048 bit RSA integers with less than a million noisy qubits," arXiv:2505.15917 (2025).

[39] Google Quantum AI and Collaborators, "Quantum error correction below the surface code threshold," *Nature* 638, 920–926 (2025). DOI 10.1038/s41586-024-08449-y. https://www.nature.com/articles/s41586-024-08449-y — the source both for the Willow coherence figures in §14 and the below-threshold result in §16.

genuinely below threshold. In July 2026 *Nature* published work by Sivak, Morvan and collaborators applying reinforcement learning to over a thousand control parameters, allowing the processor to recalibrate from its own error-detection events while computing rather than pausing for tune-up passes; this improved the surface-code result to a record $7.72 \times 10^{-4}$ per cycle and improved stability against injected drift 3.5-fold. **[V]**[40] Google has not published a two-logical-qubit gate on superconducting hardware; that milestone remains open.

**Quantinuum** reported in February 2026 up to 94 logical qubits in distance-2 error-detecting codes and up to 48 in distance-4 error-correcting codes on the 98-qubit Helios. **[V]**[41] The results are genuine and the fidelities are impressive. The caveat vendor messaging omits is that they are **postselected**: acceptance rates run as low as 25% for the 94-qubit GHZ state and 3.2% for the deepest circuits, meaning more than 96% of shots are discarded in the headline runs.

**Harvard, MIT and QuEra** published what is arguably the strongest single error-correction result of the period in November 2025: a fault-tolerant neutral-atom architecture on 448 atoms, with distance-5 surface codes, below-threshold performance demonstrated at a factor of 2.14, transversal gates, lattice surgery, and up to 96 simultaneously active distance-4 logical qubits. **[V]**[42] The authors state plainly that they remain roughly a factor of two below key thresholds for large-scale fault tolerance.

**IBM** has bet against the surface code in favour of qLDPC codes, whose $[[144, 12, 12]]$ "gross code" encodes 12 logical qubits into 144 data qubits plus 144 syndrome-check qubits and promises roughly 10× lower overhead than the surface code. **[V]** In November 2025 IBM announced Nighthawk (120 qubits, aimed at near-term advantage) and Loon (an experimental chip validating the c-couplers, six-way connectivity and reset gadgets that qLDPC requires), together with an FPGA decoder processing syndromes in under 480 nanoseconds. **[R]** As of August 2026 no demonstration of an encoded qLDPC logical qubit outperforming its physical constituents on IBM hardware has been published; the roadmap targets fault tolerance with the Starling system in 2029. **[P]**

**The honest summary [A].** Error correction crossed from theory into demonstrated engineering between 2024 and 2026. That is a real and substantial change, and it is the reason this field looks different than it did five years ago. But no machine yet performs a useful computation that a classical computer cannot, and every headline logical-qubit count in the period above carries a qualifier — postselected, error-detecting rather than correcting, or a memory demonstration rather than a computation.

[40] V. Sivak, A. Morvan et al. (Google Quantum AI and Google DeepMind), "Reinforcement learning control of quantum error correction," *Nature* 655, 879–884 (2026). DOI 10.1038/s41586-026-10759-2. Preprint: arXiv:2511.08493.

[41] N. Dasu et al., "Computing with many encoded logical qubits beyond break-even," arXiv:2602.22211 (2026).

[42] D. Bluvstein et al., "A fault-tolerant neutral-atom architecture for universal quantum computation," *Nature* 649(8095), 39–46 (2026). DOI 10.1038/s41586-025-09848-5. Publisher correction: *Nature* 650, E3 (2026).

## Part IV — The Multiverse Question

### 17. The claim

The question that motivates this section is frequently put in something like the following form: if a quantum computer explores many possibilities simultaneously, and our universe can only contain one of them, then aren't the others happening somewhere else — in parallel universes?

The most rigorous version of this argument belongs to David Deutsch, who is not a popularizer but the physicist who invented the field. His 1985 paper defining the universal quantum computer states the position in its abstract: the properties of quantum computation place an intolerable strain on all interpretations of quantum theory other than Everett's. **[V]**[43]

His most-quoted formulation appeared in *The Fabric of Reality* (1997). Deutsch challenges anyone holding a single-universe view to explain how Shor's algorithm works — not to predict that it will work, but to explain it. When the algorithm has factorized a number using vastly more computational resources than can be seen to be present, he asks where the number was factorized, given that the visible universe contains only about $10^{80}$ atoms and the computation appears to require something on the order of $10^{500}$ times the visible resources. If the visible universe were the whole of physical reality, he argues, physical reality would not remotely contain what the calculation consumed. **[V]**[44]

This is a serious argument from a serious physicist, and it deserves a serious response rather than dismissal.

### 18. The first thing to notice: it is not a prediction

Deutsch's argument is explicitly about explanatory adequacy, not empirical prediction. He says so himself in the 1985 paper: the explanations could always be translated into the conventional interpretation, though in his view not without losing their explanatory power entirely.

This concession is decisive for how the argument should be evaluated. Everyone agrees on what the quantum computer will output. The mathematics of quantum computing — unitary evolution plus the Born rule — is shared by every mainstream interpretation. Copenhagen, Many-Worlds, Bohmian mechanics, QBism, relational quantum mechanics and consistent histories all predict identical results for every quantum algorithm ever run.

[43] D. Deutsch, "Quantum theory, the Church–Turing principle and the universal quantum computer," *Proc. R. Soc. Lond. A* 400, 97–117 (1985). https://www.daviddeutsch.org.uk/wp-content/deutsch85.pdf

[44] D. Deutsch, *The Fabric of Reality* (Penguin, 1997), Ch. 9. Paraphrased; see Notes on sources.

This is not speculation. Explicit worked-out accounts of quantum computation exist in non-Everettian frameworks: Philipp Roser has given a de Broglie–Bohm (pilot-wave) treatment concluding that pilot-wave theory accounts perfectly well for the phenomena involved in quantum computing, with the speedup attributed to nonlocal guidance dynamics in a single world. **[V]**[45]

So the question is not "which interpretation does the evidence support?" — the evidence is silent. The question is "which account explains best?", and reasonable people differ about what makes an explanation good.

## 19. Five technical objections

That said, the specific "$10^{500}$ parallel computations" imagery faces serious technical difficulties — several of which are accepted by physicists sympathetic to Many-Worlds.

### 19.1 Interference is the failure of branches to be separate

This is the sharpest objection, and it comes from Scott Aaronson, who is himself a Many-Worlds sympathiser. His point is that to whatever extent different branches of the multiverse can usefully interfere for the purposes of quantum computing, to that same extent they do not behave like separate branches at all — the whole function of interference being to mix branches together until they lose their individual identities. **[V]**[46]

The argument has real force. The parallel-universe picture requires branches that are separate enough to count as universes. But the computational mechanism requires branches that are coherent — able to interfere. These are contradictory requirements. Decoherence, which is what makes branches behave like autonomous worlds, is precisely what a quantum computer engineer works hardest to prevent. During the coherent portion of a quantum computation, there are no decohered worlds to do any computing.

### 19.2 Branch counting is not well-defined

The figure $10^{500}$ counts amplitudes in the computational basis. But on the leading modern Everettian account — David Wallace's — there is no fact of the matter about how many branches there are. Branches are emergent, decoherence-defined structures; the description can be refined further and further, but there is no finest grain, because below a certain resolution interference between terms becomes significant and the branching structure disappears. **[V]**[47]

---

[45] P. Roser, "Quantum Computing in the de Broglie-Bohm Pilot-Wave Picture," arXiv:1012.4843 (2010).

[46] S. Aaronson, PHYS771 Lecture 10: Quantum Computing. https://www.scottaaronson.com/democritus/lec10.html

[47] D. Wallace, *The Emergent Multiverse* (Oxford University Press, 2012). Characterisation of branch indefiniteness per P. J. Lewis's review, *Notre Dame Philosophical Reviews*, 6 April 2013.

This creates an internal tension in Deutsch's challenge. The number $10^{500}$ requires a definite branch count in a definite basis, at exactly the stage of the computation when — by the leading Everettian's own account — no such count exists.

Adrian Kent pressed the basis-dependence point in 1990: any physical interpretation stated in one particular basis must define that preferred basis by an axiom. **[V]**[48]

### 19.3 Gottesman-Knill: branchy circuits that a laptop can simulate

Recall §12. There exist quantum circuits that generate enormous entanglement across the full $2^n$-dimensional state space — as "branchy" as anything in quantum computing — and which are efficiently simulable on a classical computer.

Michael Cuffaro has developed this into a direct argument: if branch count were what produced the speedup, stabilizer circuits would be fast, and they are not. **[V]**[49] Whatever the computational resource is, it is not the number of worlds.

### 19.4 The BBBV bound is inconsistent with independent parallel searchers

If $2^n$ branches were genuinely running independent classical computations, unstructured search would take one step. It provably takes $2^{n/2}$. Jacob Barandes puts the point well: if these universes really are there, one might expect speedups in far more circumstances than we actually observe, and it is striking that speedups appear only under very special conditions. **[V]**[50]

In fairness to the Everettian, this objection defeats the popularization rather than the interpretation. Many-Worlds predicts whatever unitary quantum mechanics predicts, including Grover's merely-quadratic speedup. It is not evidence against Many-Worlds. It is evidence against the picture of many worlds as parallel classical computers.

### 19.5 Only $n$ bits ever come out

Andrew Steane made this argument in a paper titled, pointedly, "A quantum computer only needs one universe." **[V]**[51] An $n$-qubit register manipulates only $O(n)$ bits of extractable information per step, by Holevo's bound. The $2^n$ amplitudes are not $2^n$ computations in any operational sense, because no procedure can ever retrieve $2^n$ results.

[48] A. Kent, "Against Many-Worlds Interpretations," *Int. J. Mod. Phys. A* 5, 1745–1762 (1990). https://arxiv.org/abs/gr-qc/9703089

[49] M. E. Cuffaro, "On the Significance of the Gottesman-Knill Theorem," *British Journal for the Philosophy of Science* 68(1), 91–121 (2017).

[50] J. Barandes, in dialogue with S. Aaronson, *Theories of Everything*, 4 March 2025.

[51] A. M. Steane, "A quantum computer only needs one universe," *Stud. Hist. Phil. Mod. Phys.* 34(3), 469–478 (2003). https://arxiv.org/abs/quant-ph/0003084

## 20. Deutsch himself abandoned the picture

This is not widely known and it matters more than any of the five objections above, because it comes from the argument's own author.

In his 2010 contribution to *Many Worlds?*, Deutsch writes that a collection of classical universes — even interacting ones — is still classical, and therefore has no entanglement and no phases attaching to its amplitudes; the multiverse, on his account, is not classical and consists of much more than universes. **[V]**[52]

The popular version of Deutsch's argument — $10^{500}$ parallel classical computers, each trying one factor — is a picture Deutsch explicitly repudiates. Whatever the multiverse is on his account, it is not that.

**Assessment [A].** This is the single most useful fact in Part IV, and it is almost entirely absent from popular coverage. The version of the multiverse argument that circulates in documentaries and product marketing is not a simplification of Deutsch's position. It is a position Deutsch published a rejection of, sixteen years ago, in a volume dedicated to the subject.

## 21. Where physicists actually stand

There is no consensus, and no trend toward one.

The largest survey ever conducted on quantum foundations was published in *Nature* in August 2025, with over 1,100 respondents drawn from researchers with recent quantum-mechanics publications: **[V]**[53]

| Interpretation | Share |
|---|---|
| Copenhagen | 36% |
| Epistemic views (incl. QBism) | ~17% |
| Many-Worlds / Everett | 15% |
| de Broglie–Bohm | 7% |
| Relational | 4% |
| Spontaneous collapse | 4% |

[52] D. Deutsch, "Apart from Universes," in S. Saunders, J. Barrett, A. Kent and D. Wallace (eds.), *Many Worlds? Everett, Quantum Theory, and Reality* (Oxford University Press, 2010), Ch. 22.

[53] E. Gibney, "Physicists disagree wildly on what quantum mechanics says about reality," *Nature* 643, 1175–1179 (2025). https://www.nature.com/articles/d41586-025-02342-y

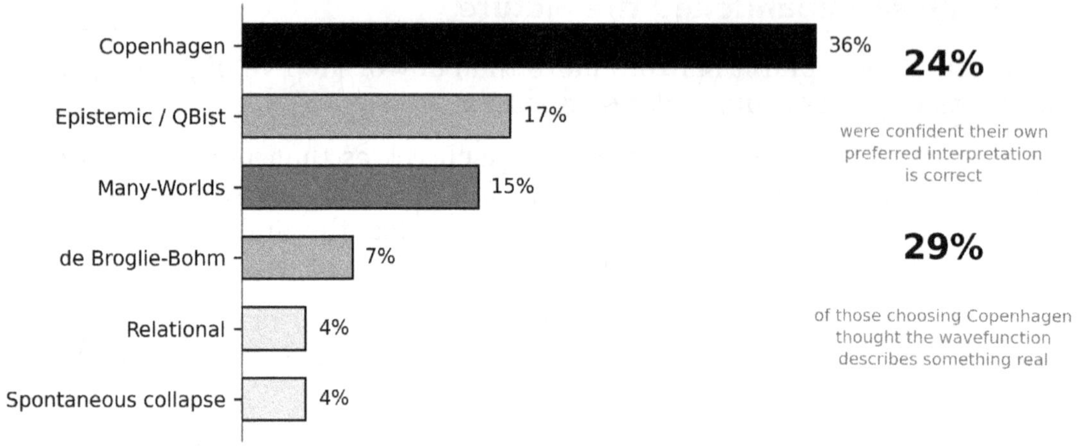


***Figure 12. What physicists actually think.*** *Nature survey, August 2025, n > 1,100. A century after the theory was written down, there is no consensus on what it means. Note also that these numbers are highly venue-dependent: a 2013 poll taken at a quantum-foundations conference returned 63% for de Broglie–Bohm and zero for Everett. Such surveys measure who was in the room at least as much as what physicists believe.*

Two findings from that survey are more revealing than the headline percentages. Only **24% of respondents were confident that their own preferred interpretation is correct.** And respondents who selected the *same* interpretation frequently disagreed with each other about what it entails — of those choosing Copenhagen, only 29% thought the wavefunction describes something real. Renato Renner of ETH Zürich described the internal incoherence as a considerable surprise.

A caution on such surveys: results vary enormously with venue. A 2013 poll at a conference on quantum foundations returned 63% for de Broglie–Bohm and **0%** for Everett — almost the inverse of the *Nature* result. **[V]**[54] These polls measure who attended, at least as much as what physicists believe.

**One gap worth stating explicitly:** no survey has ever asked physicists whether quantum computing constitutes evidence for Many-Worlds. The question has not been put.

## 22. What actually is testable

Here the discussion becomes more interesting than the usual framing allows.

Most interpretations — Copenhagen, Many-Worlds, Bohmian, QBist, relational, consistent histories — share identical dynamics and identical predictions. No experiment distinguishes them, including any quantum computation.

**Objective collapse theories are different in kind.** GRW, Continuous Spontaneous Localisation and the Diósi–Penrose gravitational model are not interpretations but rival theories: they modify the Schrödinger equation, adding a stochastic

[54] T. Norsen and S. Nelson, "Yet Another Snapshot of Foundational Attitudes Toward Quantum Mechanics," arXiv:1306.4646 (2013).

process that physically destroys superpositions above some scale. Because they change the dynamics, they make different predictions — spontaneous X-ray emission, anomalous heating, decoherence that saturates rather than growing.

And they are being ruled out. In 2021 an underground experiment at Gran Sasso searching for the predicted spontaneous radiation concluded that its result excludes the natural parameter-free version of the Diósi–Penrose model. **[V]**[55] The Majorana Demonstrator collaboration improved constraints on CSL by a further factor of 40–100 in 2022. **[V]**[56]

Now the genuinely interesting inversion. A 2022 study in *Physical Review B* examined what collapse models would do to a superconducting quantum computer, and found that CSL-induced dissipation would generate quasiparticles capable of spoiling the quantum computation of practical algorithms on large devices. **[V]**[57]

This reverses the usual claim in a defensible way. A large fault-tolerant quantum computer is not evidence for many worlds. But it *is* a meaningful constraint on objective collapse — because if the wavefunction really did spontaneously collapse at that scale, the machine would not work. Every quantum computer that scales successfully narrows the parameter space available to collapse theories.

That is a real, non-hand-wavy sense in which quantum computing bears on quantum foundations. It is just not the sense usually advertised.

Two further lines of genuine research deserve mention. Deutsch's actual proposed test of Everett — from his other 1985 paper, not the computing one — involves running an artificial observer as a coherent simulation on a quantum computer, having it report that it observed a definite outcome without reporting which, then re-interfering it. **[V]**[58] David Wallace judges the proposal technically sound but far out of reach. **[V]**[59] And in 2025, researchers encoded Extended Wigner's Friend scenarios as quantum circuits, testing "Local Friendliness" assumptions in a way that does bear on the interpretation landscape — though not in the manner Deutsch's challenge suggests. **[V]**[60]

## 23. The honest bottom line

### What can be concluded:

Quantum computers work exactly as unitary quantum mechanics plus the Born rule predicts. Every mainstream interpretation shares both. Therefore no interpretation

[55] S. Donadi et al., "Underground test of gravity-related wave function collapse," *Nature Physics* 17, 74–78 (2021). https://www.nature.com/articles/s41567-020-1008-4

[56] I. J. Arnquist et al. (Majorana Collaboration), *Phys. Rev. Lett.* 129, 080401 (2022). https://arxiv.org/abs/2202.01343

[57] M. Vischi et al., "Possible limits on superconducting quantum computers from spontaneous wave-function collapse models," *Phys. Rev. B* 106, 174506 (2022).

[58] D. Deutsch, "Quantum theory as a universal physical theory," *Int. J. Theor. Phys.* 24, 1–41 (1985).

[59] D. Wallace, retrospective on Deutsch (1985), *Int. J. Theor. Phys.* 62, 215 (2023).

[60] M. Zeng, F. Labib and V. Russo, "Towards violations of Local Friendliness with quantum computers," *Quantum* 9, 1851 (2025).

is confirmed or disconfirmed by a working quantum computer, and explicit Bohmian and Everettian accounts of quantum computation both exist and both succeed.

Deutsch's challenge is an argument about explanatory adequacy, not prediction — and he says so himself. It is legitimate to find single-world accounts explanatorily unsatisfying. It is equally legitimate not to.

The specific image of $10^{500}$ parallel computations is defective on grounds accepted across the divide: branch counts are basis-dependent and, on the leading Everettian account, not well-defined; interference is precisely the failure of branches to be autonomous; exponentially "branchy" circuits exist that a laptop can simulate; the BBBV bound contradicts independent parallel searchers; and only $n$ bits ever emerge. Deutsch abandoned the "collection of universes" picture in 2010.

Only objective-collapse theories are currently being tested. The parameter-free Diósi–Penrose model is ruled out. CSL survives in a shrinking box.

**What cannot be concluded:**

That quantum computing is evidence for many worlds. Notably, nobody who has examined this carefully endorses that inference — not Aaronson (a Many-Worlds sympathiser), not Wallace (the leading Everettian), not Clare Hewitt-Horsman (who wrote an explicitly pro-Everett account of quantum computation and disclaimed the stronger evidential claim), not Lev Vaidman (who authors the *Stanford Encyclopedia*'s Many-Worlds entry and records there that the usefulness of the Many-Worlds Interpretation in explaining the speedup of quantum computation has been questioned). **[V]**[61]

That quantum computing refutes many worlds either. The objections above defeat the popular gloss, not the interpretation.

That the question is meaningless. It has generated real experimental programmes.

The strongest defensible version of the multiverse claim is Aaronson's: Many-Worlds was heuristically fruitful. Taking the many-worlds picture seriously is what led Deutsch to conceive of quantum computing in the first place. That is a genuine and underrated form of epistemic credit — an interpretation that generates a new field of physics has earned something. But heuristic fertility is not evidence, and the two should not be confused.

## 24. Coda: a better question

The question "do quantum computers prove the multiverse?" presupposes the parallelism picture, and the parallelism picture is false. A better question, and one with a sharper answer, is this:

> What does it mean that nature computes with amplitudes rather than probabilities?

[61] L. Vaidman, "Many-Worlds Interpretation of Quantum Mechanics," *Stanford Encyclopedia of Philosophy*, revised 17 June 2026. https://plato.stanford.edu/entries/qm-manyworlds/

That is not an interpretive question. It is a fact about the world, established by experiment, and it is strange enough on its own terms. Nature keeps track of quantities that can be negative and complex, that can cancel, and whose squared magnitudes we observe as probabilities. No classical theory of any kind does this. Whether one describes the resulting structure as one world or many, the amplitudes are there, and they interfere.

Every quantum algorithm ever devised is an attempt to make that interference do something useful. It is a narrow craft — the theorems in Part II say it must be — and it is genuinely, physically new. That seems remarkable enough without the multiverse.

---

## What would have to be true

A paper on a moving field should be checkable against the future rather than merely persuasive about the present. The following are the indicators that would confirm or falsify this paper's assessments. Each is dated, and each is the kind of claim that a reader can verify without taking the author's word for anything.

**For the interference thesis (Parts I-II) — these are theorems, and would require a mathematical error to overturn.**

- If any algorithm were found giving exponential speedup on a *total* Boolean function with no promise structure, §12 would be wrong. This would contradict the Aaronson–Ben-David–Kothari–Rao–Tal result, so the more likely outcome is that no such algorithm exists.
- If a quantum algorithm solved an NP-complete problem in polynomial time, the entire framing of Part II would collapse. BBBV makes this impossible relative to a random oracle; a non-relativizing technique would be required.

**For the hardware assessment (Part III) — watch these through 2027–2029.**

- **A two-logical-qubit gate on superconducting hardware, published and not postselected.** Google has not published one as of August 2026. This is the clearest single marker that error correction has become computation rather than memory.
- **A qLDPC logical qubit demonstrated on IBM hardware outperforming its physical constituents.** IBM's 2029 Starling target depends on this and it has not yet been shown. Absence by end 2027 would suggest the surface code retains its lead despite the overhead penalty.
- **An independent, non-Microsoft replication of a topological qubit.** Fourteen years of claim-and-critique have not produced one. If Majorana 2 clears peer review and an outside group reproduces the topological phase without the topological gap protocol as an intermediary, the modality assessment in §14 changes materially.
- **Postselection rates falling below 50% in headline logical-qubit announcements.** The Quantinuum results are real; the 3.2% acceptance rate

on the deepest circuits is the honest measure of how far they are from computation.

**For the multiverse argument (Part IV).**

- **A survey that actually asks the question.** No published survey has asked physicists whether quantum computing is evidence for Many-Worlds. Until one does, claims about what "physicists think" on this specific point are unsupported.
- **Further constraints on CSL from scaling machines.** If large fault-tolerant processors operate as designed, the collapse-model parameter space narrows further. This is the one empirical channel by which quantum computing genuinely bears on foundations, and it runs in the opposite direction from the popular claim.

## Limitations

Stated plainly, so that a reader can weigh them.

**This is not a peer-reviewed paper.** It is a synthesis compiled from primary literature by an author whose professional background is operational and industrial rather than academic physics. A peer-reviewed version would require, at minimum: independent verification of every experimental figure against the original supplementary materials rather than the abstracts; a specialist referee on the foundations material in Part IV, where the philosophical literature is larger than the sample cited here; and formal treatment of the density-matrix formalism that §3.4 and §6 handle descriptively.

**Part III ages fastest.** Hardware status is current to August 2026 and this field has produced material developments monthly. Treat every number in §14 and §16 as historical if you are reading this more than twelve months after that date, and verify against the primary sources given in the notes.

**The interpretation survey is a weak instrument.** §21 relies on it heavily while also arguing that such surveys measure attendance as much as belief. Both statements are true and the tension is not resolved here.

**Selection of objections in §19 is not exhaustive.** Five were chosen for being technically load-bearing and defensible across the interpretive divide. The Everettian literature contains replies to each, and a reader persuaded by Wallace's decision-theoretic derivation of the Born rule may weigh §19.2 differently than this paper does.

**No confidential material was used.** Every source is public and cited.

## Glossary

**Amplitude** — a complex number attached to each computational path. Its squared magnitude gives the probability of the corresponding outcome (§2.1). Amplitudes can cancel; probabilities cannot.

**BBBV bound** — the 1997 result that unstructured quantum search requires $\Omega(\sqrt{N})$ queries. The theorem that makes the "tries everything at once" claim provably false (§8.1).

**Bell state** — one of four maximally entangled two-qubit states (§5.1).

**Bloch sphere** — the geometric picture of a single qubit's state space. Does not generalise to two or more qubits (§3.4).

**Born rule** — outcome probability equals the squared magnitude of the amplitude (§4.1).

**BQP** — the class of problems a quantum computer solves efficiently. Contained in PSPACE (§12).

**CHSH inequality** — the experimentally usable form of Bell's theorem (§5.2).

**Coherence** — the persistence of well-defined relative phase. What decoherence destroys (§6).

**Decoherence** — the leakage of phase information into environmental correlations. Not collapse; nothing is destroyed, only made inaccessible (§6).

**Dequantization** — showing that a claimed quantum speedup has a classical equivalent once both sides are granted the same input access (§13).

**Distance (of a code)** — roughly, how many errors a code can tolerate. Higher distance means more physical qubits per logical qubit (§15).

**Error-correcting vs error-detecting code** — a correcting code fixes errors and runs to completion; a detecting code notices errors and aborts. Press releases routinely conflate them (§15).

**Gottesman-Knill theorem** — certain massively entangled circuits are efficiently simulable classically. The decisive evidence that entanglement and state-space size are not what produce speedup (§12).

**Gross code** — IBM's $[[144, 12, 12]]$ qLDPC code, encoding 12 logical qubits with roughly 10× less overhead than the surface code (§16).

**Holevo's bound** — $n$ qubits convey at most $n$ classical bits (§4.4).

**Interference** — the addition and cancellation of amplitudes. The actual mechanism of every quantum speedup (§9).

**Kochen-Specker theorem** — definite pre-existing values cannot be assigned to all observables independently of measurement context. Requires no locality assumption (§3.2).

**Logical qubit** — an encoded qubit spread across many physical qubits (§15). The term is heavily abused.

**Monogamy of entanglement** — maximally entangled qubits cannot be entangled with anything else (§5.3).

**No-cloning theorem** — an arbitrary unknown quantum state cannot be copied (§4.3).

**Objective collapse theory** — GRW, CSL, Diósi–Penrose. Not interpretations but rival theories with different predictions, and currently being ruled out (§22).

**Postselection** — discarding runs that fail a check. Legitimate practice; misleading when the discard rate is omitted from the headline (§16).

**QCCD** — quantum charge-coupled device; the trapped-ion architecture that physically moves ions between operation zones (§14).

**qLDPC** — quantum low-density parity-check codes. IBM's alternative to the surface code (§16).

**Relative phase** — the phase difference between components of a superposition. Physical, measurable, and where the computation lives (§3.3). Contrast **global phase**, which is not physical.

**Rydberg blockade** — the mechanism used for two-qubit gates in neutral-atom processors (§14).

**Superposition** — a state that is not an eigenvector of the observable being measured. Basis-relative, and not "two values at once" (§3.1).

**Surface code** — the leading error-correcting code; threshold approximately 1% (§15).

$T_1$ / $T_2$ — energy relaxation and dephasing times. Related by $T_2 \leq 2T_1$ (§6).

**Threshold theorem** — below a critical physical error rate, arbitrarily long computation is possible with polylogarithmic overhead (§15).

**Transmon** — the superconducting qubit design used by Google and IBM (§14).

**Tsirelson bound** — the $2\sqrt{2}$ ceiling on quantum correlations (§5.2).

**Unitary** — a reversible, norm-preserving linear operation. All quantum evolution short of measurement is unitary (§2.2).

## Notes on sources and verification

This paper was assembled from primary literature where possible. Several claims carry caveats that readers should be aware of:

- The exact page reference for the Deutsch quotation in *The Fabric of Reality* could not be verified; it appears in Chapter 9. The text is corroborated across three independent secondary reproductions, and is paraphrased rather than quoted here.
- The Feynman quotation about nature not being classical, though universally attributed to his 1982 paper, could not be verified against the original journal text and is therefore not quoted.

- The Renou et al. result on real-valued quantum mechanics is contested; see §2.3.
- Two-qubit gate fidelity records circulated by vendors are excluded where no peer-reviewed source could be found.
- Where a claim originates in a company announcement rather than peer review, it is graded **[P]** or **[R]** and said so in the text.
- Hardware figures in §14 and §16 were re-verified against primary sources in August 2026. The Helios results are Nature-published and independently benchmarked by Sandia; the Majorana 2 claims are company-reported and not peer reviewed at the time of writing.
- Figures in this edition were regenerated in grayscale so that they remain legible in a black-and-white print interior. No figure distinguishes its series by colour alone.

## About the author

Kristian Magnuson is founder and principal advisor of OpX Advisory Group, an operational excellence and business transformation consultancy. He has spent more than twenty years in global leadership roles across manufacturing, distribution, supply chain, engineering and enterprise business excellence, where the recurring problem is the gap between what a technology is claimed to do and what it can be relied upon to do on a Tuesday.

That lens is what produced this paper. Quantum computing is a field in which the promotional literature and the peer-reviewed literature describe two different machines, and the discipline of separating verified fact from vendor projection transfers directly from industrial transformation work to this subject. He writes on quantum computing, advanced energy, and the operational realities of emerging technology.